Parkinson's Disease Dravidian Cure Chintharmony System

Vasu Jayaprasad

Lulu.com

Parkinson's Disease Dravidian Cure Chintarmony System

Contents

Preface

Chapter I

Origin of Chintarmony System

A. Diagnosis

B. Treatment

C. Massage

D. Mudra Kriya, the Sealing Act

E. Parkinson's Disease

F. Neck Injury

G. Vertebral Column Injury

Chapter II

Modern Medicine

A. Antioxidant Properties

B. Vitamin C Mega Dosage

C. Regulation of Vitamin C

D. Advocacy Arguments

E. Physical Exercise

F. Case Study

    G. Australian Chiropractor has achieved it!

Chapter III

Symptoms

A. Stages: Dizziness and Hunger Pain

B. Constipation

Parkinson's Disease Dravidian Cure Chintarmony System

1. Naturopathic Enema

2. Enema Cleansing and Retention

3. Examples of Cleansing Enemas

4. When Enemas Are Really Important

C. Sleep disorder

1. Herring's law of Cure

2. Physiology of Information

D. Malnutrition

Chapter IV

Logo therapy

1. Rational Emotive Behavioral Therapy

2. Cognitive Dissonance

Chapter V

Causes of Parkinson's disease

A. Multiple Sclerosis

1. Brain Blood Barrier Break Down

2. Auto Immunology

3. Lesions

4. Inflammation

B. Salt Deficiency

C. Food Culture

D. Fear Factor

E. Postures

F. Lactose Intolerance

G. Gluten Related Injury

1. Wide Ranging Injury

2. Ataxia

3. Hearing Problems

Parkinson's Disease Dravidian Cure Chintarmony System

4. Cognitive Decline

5. Expanding Horizons

6. Theories

7. Bacterial Balance

Chapter VI

Sidha Cult

Chapter VII

Basics of Sidha Medicine

1. Concept of Disease and Cause

2. Method of Diagnosis

3. Concept of Medicine

4. Concept of Treatment

5. Concept of Physician

6. Approaches

7. Varma Branch of Sidha Medicine

Chapter VIII

1. Action step

2. Conclusion

#

This book is dedicated to the memory of Justice Varghese Kalliath Former judge of the High Court of Kerala, who died recently suffering from Parkinson's disease well known for his judgment in Ammonia storage litigation in

AIR1994 Kerala Page308 Law society of India Vs Fertilizers and Chemicals Travancore Ltd. Justice Kalliath was hit at TDLH value when he visited

ammonia storage in Willingdon Island.

Parkinson's Disease Dravidian Cure Chintarmony System

PREFACE

Though modern medical research establishes that Parkinson's disease is not caused by damage to substantia nigrum, the outer covering of brain impairing ability of brain to produce dopamine neuron, doctors promote dopamine drugs. The Patients are demoralized to make them surrender to exploitative medical malpractice. Author attempts to give confidence to such patients to look around to lead a fulfilling life.

Author introduces one of the oldest systems of medicine dating 5000 B.C that is practiced even now by Dravidian descendants now settled in southern India. Chintarmony system of treatment now is on the verge of extinction offers accuracy, very short duration of treatment and predictable cure. The system functions based on removal of blocks in the energy points. Thai Chi, pranic healing, reiki and Kalari medicines of Kerala have close resemblances to this medical practice. Modern medicine and scanning technology trail much behind Dravidian cure.

MW151 andMw189 are the new panacea offered. Studies show it introduces auto immune disorders. A holistic approach is made around 5000 years back. Do we have to invent wheel again? It is like a drunkard may take to drinks to dehydrate and stifle his shaking hands!

Kochi

20th June2012　　　　　　　　　　**Vasu Jayaprasad**

Parkinson's Disease Dravidian Cure Chintarmony System

Chapter 1

ORIGIN OF CHINTARMONY SYSTEM

From

Dravidian Cure

Bhojar Sidha was one of the eight disciples of Agastya Sidha who founded the system of medicine. Agastya's tomb is Anandapadmanabha temple, the richest temple in India with 10,000 billion USD worth gold precious, stones and jewellery. Sidha's are highly regarded for their learning in medical skills. They have a perception of super science, which is inscrutable even from the angle of modern technology. Their skill in effecting cure for many health conditions like stroke to central nervous system, multiple sclerosis Parkinson's disease is scientifically proved and accepted though process of treatment or cure defies present level of information to the modern science. The scientific community has a tendency to treat the system as occult science in the inability of modern science to understand the system and plagiarise it. Patients are often impressed by the predictable and accurate cure. Rigor of taboos insisted by the physician are sometimes annoying but scientifically justified.

Bhoja Sidha's tomb is Palani temple in Tamil Nadu, India where pilgrims worship Lord Subramanya for learning. Offerings from the temple are believed to heal and cure neurological debilities of devotees. Bhoja evolved a cure using electrical circuit of the body. Electro homoeopathy popular in Italy may be an adaptation on this system.

The cure is protected and guarded by family tradition and is prevalent in China and India. In China treatment is more associated with Thai Chi martial art and injuries from combat. Thai and Chiva represent Goddess Parvathy and god Siva, positive and negative charges in the human body. Dravidian flow in to China and India is attributed to the movement of African geotectonic plate. Yet another version points to Dravidian migration from Africa to China and into India and the cure is found predominantly among Tamil settlers in China and India.

Parkinson's Disease Dravidian Cure Chintarmony System

A. Diagnosis

By reading the pulse points in the palm with fingers, the Chintarmony specialist identifies exact point where electrical fault causes body imbalance. Injured nerves are identified from the group of healthy nerves with unbelievable accuracy in seconds. They are able to predict impending stroke better than any scanner. Diseases like Parkinson's disease, diabetes, epilepsy and stroke related physical infirmities are often cured in predictable time by removing energy blocks from ten vital points. As proven cure is not available in modern system for the above maladies, practice of chintarmony system is legal.

B. Treatment

Medicines are decoction of oils and powders sourced from exotic and rare plants taken without destroying electrical charges i.e. anti oxidant properties of plant derivatives. Metal knives or pots are not used to preserve chemical composition, electrons and charge of medicinal plants. Collectors of plant material also strictly observe vegetarian diet. Electrons that are intended to remove free radicals formed in the cells are not lost during the production of the medicine. It is more like using a new electric cell and the used one. Chemicals in the new cell are entirely different from the used cell that has discharged electricity. Sidhas are alchemists known for their expertise in very powerful medicines and uses such medicines accurately. One has to be careful in choosing a practitioner as many immature minds have entered the field as it is maintained as a family tradition. Though formal Sidha institutions are established to standardise the cure, real expertise and cure still exists outside the campus and is rigidly protected.

Modern medicine is still reinventing the wheel in spending multi billion dollars in research to find a medicine for Parkinson's disease. Magical remedies act of India will not apply to any one offering cure to Parkinson's patient, as there is no known cure in allopathic system.

B. MASSAGE

Massage with medicated oil with antioxidant properties varies from structured days like 3, 5, 11 or more depending on the serious nature of the injury. Equal days or more are spent as rest to enable the medicine to spread in the body. Strict adherence to discipline is insisted as the treatment makes the body very tender and vulnerable to injury. Often patients are isolated from surroundings to ensure stress free rest and complete cure.

Author has personally observed an acute Parkinson's patient who was abused with dopamine for around 15 years. After the administration of the medicine the patient reported that he gained flexibility in the neck. The stiffness encountered by the patient was lost in a week after oil bath and the medicine orally introduced in powder form mixed with honey. The treatment introduced new black youthful hair growth under a layer of grey hair indicating that anti oxidant action has salvaged brain damage. The patient had started limping and falling, a sign that the patient is preparing for a long and final bed rest. The physician was pleased with the development and assured complete cure after three days massage. As the patient had delayed the treatment for long, the physician recommended medication and medicated oil bath to continue for 90 days. The patient also has to repeat oil massage with the assistance of the physician for 5 days and 14 days after a break. He was hopeful of assuring complete cure reading the pulse of the patient and monitoring progress so far achieved. The dedication and commitment to the treatment is considered a divine duty and function by Chintarmony practitioners.

C. MUDRA KRIYA, THE SEALING ACT

Mudra kriya is a special massage using three fingers to seal the chakras after massage. The special oil activates nerve centres or chakras. Only expert masseur is allowed to use special oils Guru thailam (master oil). The apprentices are allowed to use only lower form of oils to avoid perpetual injury to the marmas of the body. Such injuries can be much damaging.

Many stages of Parkinson's diseases refer to oxidation (vatha) levels. Chintarmony system refers to 10 vatha conditions. Final stage refers to 10 vath level or dasa vatha affecting upper region of vertebral column causing shrinkage of brain making the patient to slip and fall. The patient then reaches a point of no return. Approach of Chintarmony system corresponds to the stages charted in modern medicine. When vatha or oxidation hits brain the insulation of nerves get destroyed and may make it difficult for Chintarmony physician for an effective cure. In the final stage the patient develops intense multiple sclerosis a condition described as incurable by modern medicine. Patient develops unsteady limb movements; Author has experienced Chintarmony system curing such patients where the vatha has damaged the brain.

D. PARKINSON'S DISEASE

Chintarmony physician only terms injured nerve in the head leading to substantia nigrum as Parkinson disease. Only 6 days is often taken to massage the head and heal the injured nerve with medicated oil and electrical charge flowing from the body of the physician. The process resembles reiki or pranic healing. Malnutrition caused by injured nerves in the C5 and C6 vertebral column of the neck or injured nerves in the vertebral column below neck is easily corrected by using oil with anti oxidant properties and massage of nerves and nerve centers. Mudra kriya is an act with three fingers to manipulate and activate nerve centers. Parkinson's disease is being renamed on the basis of symptoms, as substantia nigrum damage theory is proved wrong.

E. NECK INJURY

The injuries of vertebral column in the neck region from C1 to C7 can induce various health conditions adversely affecting internal organs from shoulder to thighs. The treatment involves whole body massage and decoctions. The physician will not delegate the massage to any apprentices. Injury to 5th and 6th vertebral column damages the digestive system. Even if the patient suffers digestive pangs and eats well, the food consumed will not be absorbed. The patient suffers from lack of R.E.M. sleep preventing dopamine neuro hormone production causing faulty movement of limbs. Body shakes and shivers like a starved person. Autopsy conducted on the Parkinson's patients disprove the approach of medical practitioners. None of the patient's brain showed signs of damage to substratum nigrum. Chintarmony system on the other hand describes the condition as inability of dopamine stocked in the brain to break the brain barrier and enter the blood stream. Rapid eye movement sleep is generally disturbed in Parkinson patients. Acute constipation and disturbed sleep are common trait of Parkinson patient.

F. VERTEBRAL COLUMN NERVE INJURY

Vertebral column injury in the abdominal area may damage digestive system due to injured nerves. Parkinson's disease like injury is caused due to lack of absorption of nutrients and lack of

Parkinson's Disease Dravidian Cure Chintarmony System

R.E.M sleep. Treatment involves whole body massage with medicated oils and decoctions for internal administration. Around 5 to 11 days are often needed for massage and equivalent days are given for rest. As minute nerves are affected by oxidative stress after the massage due to formation of free radicals, strict adherence to instruction is insisted. When free radicals drained with anti oxidants any form of exercise causes further oxidative stress neutralizing anti oxidants. Reading is prohibited to prevent injury to eyes. Modern sports medicine also validates this approach of Chintharmony system.

Parkinson's Disease Dravidian Cure Chintarmony System

Chapter II

Modern Medicine

Modern medicine admits their limitations and asserts that Parkinson Disease is incurable in their system. Medical practice and theory are conflicting. Treatment with administration of Dopamine only admittedly manages the patient with disastrous reactions. Treatment is counter productive. The morale of the patient is destroyed. Autopsy conducted on Parkinson's patients reveal that not even one patient has suffered injury of the substantia nigra. Research establishes that patients have suffered sleep disorder, stiff neck and stomach disorder due to accumulated stress. Most of the patients have suffered PD after shifting to wheat diet due to gluten allergy. Dopamine administration may be an easy way to disaster. The approach is fundamentally wrong and misleading.

Dopamine neuro hormone is extracted from a tropical plant popularly known as naykurna (macuna pruriens). Dopamine releases peroxides and patient suffers oxidative stress due to accumulation of free radicals that can cause serious injury to patients. Dravidian pharmacopeia has a purification process. If the purification process is strictly adhered, dopamine will not harm patients.

The seed is boiled in milk. Toxin contained in the milk is decanted and removed and the seeds are sun dried. The process is repeated seven times. The pulverized seed is given to patients in milk. That is considered a healthy drink, which nourishes the cells. Though modern pharmacopeia borrowed macuna pruriens from the Dravidian system the copy cat has not done a good work!

Think twice before touching dopamine in treating Parkinson's patient or statin tablets for cholesterol related problems. When doctors speak diplomatically and equivocally like a Jesuit or Dharmaputra (character in great Indian epic Mahabharata misunderstood widely as an ideal character) double check them. Your life could be in danger. Lipitor is the single largest selling medicine after 1987. About I35$ billion dollar is said to be yearly turn over. This feat is achieved by well-orchestrated propaganda about bad cholesterol and good cholesterol. Ldl is taught in medical schools as a little rascal and not a saviour in medical schools.ref rath research foundation. In Kerala even a child knows statin from chilly dilutes the

Parkinson's Disease Dravidian Cure Chintarmony System

blood and are prohibited from consuming excess chilly in the diet. There is volume of scientific proof that medicines like statin, dopamine drugs injures the patient seriously and were never offered as proven cure.

It is also scientifically proved that angioplasty or biopsy has not cured any patient. On the other hand alternative medicines and traditional herbal medicines have offered scientifically proven remedy though practitioners do not have scientific knowledge about how the cure is achieved, with the scientific jargon available with the modern medicine.

If a patient in America consume statin tablets and develops body pain, he need only call a toll free number and report that the patient consumed poison and need medical assistance. Ambulance service is immediately available. Right remedy is perhaps introducing heavy doses of vitamin C. Parkinson's disease according to Chintarmony system is oxidation related. Vitamin C supplies electrons internally as anti oxidants.

A.   Anti oxidant property

Parkinson's disease is considered as related to vatha or oxidation of cells due to release of free radicals destroying electrons in the cells. Cure is by applying medicated oil containing electrons, which are anti oxidants. Decoctions with anti oxidant electrons are orally given to rejuvenate cells and contain cell damage. During the treatment patient is curbed from all forms of exercise including yoga.

An antioxidant is a molecule capable of inhibiting the oxidation of other molecules. Oxidation is a chemical reaction that transfers electrons or hydrogen from a substance to an oxidising agent. Oxidation reactions can produce free radicals. In turn, these radicals can start chain reactions. When the chain reaction occurs in a cell, it can cause damage or death to the cell. Antioxidants terminate these chain reactions by removing free radical intermediates, and inhibit other oxidation reactions. They do this by being oxidized themselves, so antioxidants are often reducing agents such as thiols, ascorbic acid, or polyphenols.

Although oxidation reactions are crucial for life, they can also be damaging; plants and animals maintain complex systems of multiple types of antioxidants, such as glutathione, vitamin C, and vitamine E as well as enzymes such as catalyse , superoxide dismutatase and various peroxidases. Low levels of antioxidants, or inhibition of the antioxidant enzymes, cause oxidative stress and may damage or kill cells.

As oxidative stress appears to be an important part of many human diseases, the use of antioxidants in pharmacology is intensively studied, particularly as treatments for stroke and neurodegenerative diseases. Moreover, oxidative stress is both the cause and the consequence of disease.

Antioxidants are widely used in dietary supplements and have been investigated for the prevention of diseases such as cancer, coronary heart disease, coronary heart disease and even altitude sickness. Some studies conducted by Linus Pauling Nobel Laureate proved that Vitamin C is helpful in the cure of Coronary blocks without bye pass or angioplasty (reference rath researce foundation org).Vitamin C has a curative effect on cancer patients and auto immune disorders. Parkinson patient exhibits the stooping gate of patients with auto immune disorders, epilepsy or multiple sclerosis.

A. Vitamin C Mega Dosage

Mega dosage is the consumption (or injection) of vitamin C (ascorbic acid) in doses comparable to the amounts produced by the livers of most other mammals and well beyond the current Dietary Reference Intake. A practitioner of vitamin C mega dosage might consume several grams or commonly up to 20 grams or more per day under the assumption it leads to optimal health or healing of some condition. The dosage is usually divided and consumed in portions over the day. Injections of hundreds of grams per day are advocated by some physicians for therapy of certain conditions, poisonings, and recovery from trauma. People who practice vitamin C mega dosage may consume many vitamin C pills throughout each day or dissolve pure vitamin C crystals in water or juice and drink it throughout the day.

Intravenous application of vitamin C around 50gms per day salvaged a patient in Kochi from dialysis, lowering creotinin and completely restoring kidney function. Intravenous application of mercury is referred in Araku nool sastra written by Ravana in Tamil language. Ravana and are referred in epic Ramayana. Vitamin C mega dosage is yet to be tried as an anti oxidant to Parkinson's patient. Mercury is considered as source of energy and was applied with absolute accuracy. Chintharmony system commands absolute accuracy and care.

Parkinson's Disease Dravidian Cure Chintarmony System

A. **Regulation Of Vitamin C**

There are regulations in most countries which limit the claims on the treatment of disease that can be placed on food, drug, and nutrient product labels. Regulations include:

- Claims of therapeutic effect with respect to the treatment of any medical condition or disease are prohibited by the Food and Drug Administration in the USA, and by the corresponding regulatory agencies in other countries, unless the substance has gone through a well established clinical trial with neutral oversight.
- In the United States, the following notice is mandatory on food, drug, and nutrient product labels which make health claims:

These statements have not been evaluated by the Food and Drug Administration. This product is not intended to diagnose, treat, cure or prevent any disease.

D. **ADVOCACY ARGUMENTS**

Vitamin C advocates argue that there is a large body of scientific evidence that the vitamin has a wide range of health and therapeutic benefits but that this belief is rejected by current science and medical research.

There is some evidence regarding the applications and efficacy of vitamin C, but recommended governmental agency doses and frequency of intake have remained relatively fixed. This has led some researchers to challenge the recommendations. In 2003, Steve Hickey and Hilary Roberts of the Manchester Metropolitan University published a fundamental criticism of the approach taken to fix the nutritional requirement of vitamin C. In 2004, they again argued that the RDA, which is based on blood plasma and white blood cell saturation data from the National Institutes of Health (NIH), was based on flawed data.

According to these authors, the doses required to achieve blood, tissue and body "saturation" are much larger than previously believed. Oral intake of gooseberry juice with aloe vera juice has corrected malnutrition in bed ridden parkinson's patient with shrunken brain gaining mobility from some studies in Kochi.

B. **PHYSICAL EXERCISE**

Parkinson's Disease Dravidian Cure Chintarmony System

During exercise, oxygen consumption can increase by a factor of more than 10. This leads to a large increase in the production of oxidants and results in damage that contributes to muscular fatigue during and after exercise. The inflammatory response that occurs after strenuous exercise is also associated with oxidative stress, especially in the 24 hours after an exercise session. The immune system response to the damage done by exercise peaks 2 to 7 days after exercise, which is the period during which most of the adaptation that leads to greater fitness, occurs. During this process, free radicals are produced by neutrophilis to remove damaged tissue. As a result, excessive antioxidant levels may inhibit recovery and adaptation mechanisms. Antioxidant supplements may also prevent any of the health gains that normally come from exercise, such as increased insulin. Ancient Chintarmoni system works on this simple logic. Stringent restrictions are not liked by the patients who rush to unscientific quickies to get instant relief offered by modern medicine. Modern research also supports the wisdom of ancient sages.

The evidence for benefits from antioxidant supplementation in vigorous exercise is mixed. There is strong evidence that one of the adaptations resulting from exercise is a strengthening of the body's antioxidant defenses, particularly the glutathione system, to regulate the increased oxidative stress. This effect may be to some extent protective against diseases which are associated with oxidative stress, which would provide a partial explanation for the lower incidence of major diseases and better health of those who undertake regular exercise.

No benefits for physical performance to athletes are seen with vitamin E supplementation. Despite its key role in preventing lipid membrane peroxidation, 6 weeks of vitamin E supplementation had no effect on muscle damage in ultra marathon runners. Although there appears to be no increased requirement for vitamin C in athletes, there is some evidence that vitamin C supplementation increased the amount of intense exercise that can be done and vitamin C supplementation before strenuous exercise may reduce the amount of muscle damage. Other studies found no such effects, and some research suggests that supplementation with amounts as high as 1000 mg inhibits recovery.

A review published in Sports Medicine looked at 150 studies on antioxidant supplementation during exercise. The review found that even studies that found a reduction in oxidative stress failed to demonstrate benefits to performance or prevention of muscle damage. Some studies

Parkinson's Disease Dravidian Cure Chintarmony System

indicated that antioxidant supplementation could work against the cardiovascular benefits of exercise.

In Chintarmony treatment complete rest is insisted for patients as pathym (curbing food life style liked by the disease for the aggravation). Even Yoga exercise which increases oxidation level is tabooed as that obstructs antioxidants to work without interruption. Excercises and yoga act like walking in to a room and soiling the area when the floor is being cleaned with water. Cleaning process gets obstructed and cells will not be able rejuvenate fully. To prevent physicians from experimenting with patients they are expected to follow blindly and rigorously the treatment without understanding the super science behind the structured treatment. A conscious gap is created between practioners of medicine and researchers to derive maximum advantage to the patent without financial motives. Present day also there is a conscious gap between research finding and actual medical practice with ulterior financial motives and advantage to the medical business. In the past commercial interest was delinked from patient care.

F.   Case Study By A Non Doctor Mr. G.A. Mathew

Sir,

Thanks for the encouraging words. They amply compensate for the dreary miles travelled, precious dollars spent and hours of reading and study.

I shall send the books as desired.

I am delighted to learn that at least you have cared to understand how the Rath foundation is lagging behind while some others have forged ahead valiantly with efforts to provide the high doses, braving odds from the mainstream medicine and its henchmen. It is largely because of this snail-paced approach of the Rath Foundation that I did not care to attend the functions in Kochi.

Sorry for this long-winded letter but I am restless until I share an experience with you. Alphonsa and I have been married for 4o years. We never ever spent any money on treatments. But last September, she got a nasty cut on her thumb from the mixie grinder. She was given a fairly high dose of antibiotics by the doctor for a disconcertingly long period.

Parkinson's Disease Dravidian Cure Chintarmony System

Subsequently she developed fatigue and lack of appetite. For a few days she refused to get medical attention, and when she did go to the doctor she was diagnosed as having urinary infection. It was a small clinic in Angamaly. The doctor was very confident of tackling it with the usual antibiotics. But a few days' treatment made her worse. Then she was admitted to the biggest hospital in the area. She underwent all those usual tests and diagnostic procedures. She spent about a month in the hospital, treated apparently by the best of physicians.

The tragedy was, no antibiotic could inhibit the virulent E-coli bacteria. They had become resistant. Her infection remained unabated, her hemoglobin plummeted and creatinine soared.

Doctors decided on blood transfusions and dialysis. But I knew that until the infection was controlled nothing could benefit. My son (highly qualified doctor in Australia, an F.R.C.S.)also insisted on dialysis and blood transfusions. But I pleaded with the doctors to delay the drastic measures just for three days. Meanwhile under the veil of secrecy I administered vitamin C powder in apple juice, and promptly the kidney function remarkably improved. Creatinine came down to 1.8 from 4.8. Shame on doctors who pontificate that vitamin C will knock out the kidneys!

In the case sheet the doctor had already marked renal failure which I hid from my wife, not to scare the hell out of her.

But the infection refused to budge a tad, and the hemoglobin dipped alarmingly low. Every test showed unmistakable antibiotic resistance.

I knew from my direct observations in the finest alternative healing centers in the US and Mexico and from constantly updating my information with the latest research revelations that vitamin C in ADEQUATE DOSE was a powerful antibiotic and antiviral and that AT ANY HIGH DOSE IT WAS AS SAFE AS MILK. I knew of doctors who easily tamed aggressive infections with vitamin C. Both IV and frequent oral doses together with other nutrients would be a must.Survival of a snake despite repeated strokes does not mean that it can defy death but that the strokes were not hard enough. It was clear that my cladestine servings of vitamin C were not adequate. It did n't mean that vitamin C was ineffective.From my earlier internet explorations, I knew of two suppliers of vitamin C IV fluid. One was in Coimbatore and the other one in Madurai. The Coimbatore supplier had 25 gram bottles and the madurai man had 5

Parkinson's Disease Dravidian Cure Chintarmony System

gram bottles. Incidentally, they were meant for microgram level administrations, like one gram each for a patient, to be included in chealation treatments. A few yers ago, I had pleaded with a provider of such treatment to give me a five gram infusion of C, and he trembled like an aspen leaf! Nobody knew of such infusions.

My wife was slowly but almost certainly edging towards the brink and I got her eased out of the hospital .I promptly dashed to Coimbatore (I kept everyone, including my driver, out of the picture) and returned with ample supply of vitamin C IV fluid.

I had two friends in Kochi, both doctors, in Government service, who had earlier visited me and lavished accolades and presents, in good appreciation of my books. I requested them to come immediately to administer the IV which they did. She was given 50 grams as IV. The next day I pleaded with the doctor in the small clinic who had treated her in vain for her infection to keep giving her the IV infusions .He complied.

She turned the corner soon after. Marked improvement came after 13 infusions and she became her former vivacious, energetic, amiable self after 30 IVs. The last 10 infusions were of 62.5 grams. She later confided to me that the sisters at the clinic had confidentially insisted that she be taken to the Lakeshore hospital as they were not very hopeful of a successful outcome with this unheard-of vitamin C protocol.

Now she is as healthy as ever. My son, together with his wife(with similar medical credentials)gleefully think that she recovered thanks to the delayed action of the antibiotics! I have never breathed the real thing to a soul, especially to the near and dear ones. Why rock the boat! May they live with their blissful ignorance!

Interestingly a couple of doctors are now toying with the idea of vitamin C infusions. They have already administered it to some cancer patients. Very soon many may appear on the horizon.

Sorry for bothering you with my personal anecdote. But is n't it revealing!

With the information doled out from the Rath Foundation she would not be with me today.

Thanks and regards,

Mathew, Manjapra

Parkinson's Disease Dravidian Cure Chintarmony System

G.   Australian Chiropractor Has Achieved It!

Australian chiropractor Noel Batten seems to have evolved a cure for Parkinson's disease through physical exercise and vertebral column correction with the assistance of a chiropractor. Chintarmony system seems to be more sensible as it addresses damaged nervous system and rejuvenating oxidized tissues. The system reconstructs cartilage also. Vertebral column correction involves possibility of injury to the nerves. By applying medicated oils and medicines for softening the muscular tissues the possibility of snapping of the nerves accidently is prevented. Also exercise induces further oxidation of cells leading to accumulation of free radicals destroying the cells. Chintarmony system insists on patient to bring down oxidation rate by cutting out all physical activities including yoga excercise. The system seems to be more advanced though evolved 5000years ago.

Anti oxidants are given externally as oils and internally as herbal decoctions. The patient responds positively to the application of medicine immediately. Often patient gets fast relief and is amazed.

#

Parkinson's Disease Dravidian Cure Chintarmony System

Chapter III

SYMPTOMS

Missing meals lead to dizziness and shaking of the body. The gait also changes. There is change in posture. Symptoms vary depending on nutrition, sleep and emotions suggesting substantia nigra is not damaged. A person responding to acute anger or fear responds the same way as Parkinson's patient. Nervousness, stiffness and tremors are symptoms common to patients suffering from multiple sclerosis, auto immune disorders, and Parkinson's patients. Professional approach is a process of elimination to identify exact condition

A Parkinson's patient experience the above symptoms despite the fact nourishment are there regularly. Patient suffers from malnutrition as digestion is impaired due to variety of reasons. Chintarmony system founded by Bhoja Sidha around 5000 BC made more scientific approach than the physicians of modern medicine. Reference Noel batten.

A.   Dizziness and Hunger Pain.

Parkinson's patient behaves like a famine stricken man even immediately after nourishment. He sweats and shivers and suffers intense experience of sweating and shivering as if subjected to extreme hot and cold. Shivering is characteristic. He suffers dizziness and hunger pain. Some immediately respond to herbal decoctions that correct bacterial imbalance, Gluten allergy, lactose intolerance villi damage etc.

Dopamine administration destroys the patient steadily and completely in 10 to 15 years. Dopamine is a drug administered for regulating pressure. Such medications steadily damage heart function. Patients only progresses from bad to verse. Only escape route for a patient is to equip himself in the study of comparative medicines.

B.   Constipation

Parkinson's patient suffers acute constipation and digestive disorders. That could be the reason for disturbed sleep. Disturbed sleep again causes indigestion and constipation. This is a vicious cycle. Inducing sleep is the first step in building confidence and instilling hope in to the patients psyche.

Parkinson's Disease Dravidian Cure Chintarmony System

1. Naturopathic Enema

An enema is a method to flush waste out of the colon. It's not such a bad idea when you consider that the average person may have up to 10 pounds or more of non-eliminated waste in the large intestine! We have learned many methods of healing from animals. Elephants introduced enema to human pharmacopeia. Elephants consume huge quantity of fiber in their daily life. Therefore constipation has a killer effect on them. People experience high pressure in their body when they suffer constipation. Simple way to relieve constipation is to have a simple enema kit clean the blocked rectum and address root cause later on. Parkinson patients get good sleep when they go to bed after colon cleaning enema.

Simply put, an enema cleans up the colon and induces bowel movements, leaving you feeling cleaner, lighter and healthier almost immediately.

2. Enema Cleansing and Retention

There are two types of enemas - cleansing and retention.

The cleansing enema is retained for a short period of time until your natural peristaltic movement eliminates both the water and the loose fecal material. It is used to gently flush out the colon.

The retention enema is held in the body for longer. For example, the famous "coffee enema" is retained for approximately fifteen minutes or can also be left in and absorbed. Coffee enemas are an example of short term (fifteen minutes) retention enema. They were made popular by Max Gerson who used them with cancer patients to open the bile ducts increasing bile flow, helping to rid the liver of impurities.

3. Examples of Cleansing Enemas:

- Lemon Juice - Just what you need to clean the colon of fecal matter, balance its pH and detoxify the system
- Apple cider vinegar in water - helps with viral conditions and to clear mucous from the body. Great if you suffer from nasal congestion or asthma.
- Catnip Tea - Relieves constipation and congestion and will bring down a high fever

- Burdock Root - Helps to eliminate calcium deposits and purify blood Examples of Retention Enemas:
- Coffee - A coffee solution (we mean a good organic breakfast blend, not decaf or instant) stimulates both the liver and the gallbladder to release toxins. (15 minutes only)
- Minerals - this is one you will want to retain permanently. It helps rebuild the energy of the adrenals and the thyroid
- Probiotic - Perfect for candidiasis and other yeast infections
- Red Raspberry Leaf- High in iron, great for the eyes and particularly helpful for women
- Each enema requires a slightly different method, but the results for each will be glorious. When a smaller amount of liquid is retained permanently we prefer to call this an implant. One cup of liquid with a probiotic, minerals or something green with chlorophyll (like wheat grass) makes an excellent implant. They'll quickly have you on your way to a happier, healthier colon!

4. When Enemas Are REALLY Important

Whenever you're feeling lethargic or constipated, it's a sure sign that an enema should be scheduled into your immediate future. Even if you're feeling great, releasing extra toxins can only make you feel better!

But as you approach a healthier lifestyle and begin the Body Ecology Way, home enemas are incredibly important because:

Fermented Foods - As you begin to add fermented foods and drinks to your diet, such as cultured vegetables, keifir, coco biotic or our new innergy-biotic, you might experience symptoms of yeast and pathogenic bacteria die-off as the good bacteria begin to colonize your inner-eco system. Gas and bloating is a signal that the mighty floras are busy attacking toxic waste in and on the colon walls.

Home enemas will rinse out the toxins, helping to shorten the discomfort you might be feeling.

Parkinson's Disease Dravidian Cure Chintarmony System

Fasting - If you choose to fast, enemas are critically important. You must keep things moving out of your body because your cells will be dumping their toxins quickly. You will feel miserable if you are not cleaning downstream.

Colon Cleansers - If you're using any type of clay cleansing agents to pull toxins out of the system, you'll want to use an enema to help release the bulk. You don't want to pull out toxins and leave them with nowhere to go!

Liver Cleansing - Because your liver dumps its toxins into your colon, you want to make sure the colon is open and in perfect working order to move all those toxins out quickly.

Patient suffering from constipation and evacuation problem also suffer from acute sleep disorder. Digestive disorder is one of the multitudes of factors that induce constipation. Author has designed a herbal soup to manage constipation due to villi damage in a healthy way. The soup will not give solution in cases of digestive disorder induced by vertebral column injury. Nerves injured by degenerated vertebral column can be healed perhaps only by Dravidian Cures. Acute fever managed by homeopathy sometimes act as permanent cure. Author has experienced such a cure. An American friend suffering from dislocated vertebral column C1 and C2 met a staff in income tax office in Rajastan to heal the injury in two minutes. Author feels such quick cures as risky as surgery suggested by neuro surgeon in US. These quick procedures may snap nerves permanently.

C.   Sleep Disorder

Rapid eye movement (R.E.M) sleep is necessary at least for two hours to repair and recreate damaged cells. Parkinson's patients suffer from (R.E.M) sleep is testified by medical research.

i.   Herring's Law of Cure

'Extracted from Noel Batten 'Beautiful sleep abundant energy'

'As he practiced the art of homeopathy in Europe during the eighteen hundreds, Constantine Herring proposed a law that promoted with great success, the ability of self-healing in the search

Parkinson's Disease Dravidian Cure Chintarmony System

for a personal cure, through the natural functions of the mind and body. Herring's Law of Cure reads as follows. "All curative effects come from within the spine, out to the organs, from the improvement of the mind down into the body and in the reverse order as the symptoms first established.' This is the basic concept of Chintarmony system also. System addresses blocks in the central nervous system flowing from vertebral column towards brain and to the limbs. Treatment begins from restoring brain by rejuvenating brain, vertebral column, limbs, internal organs and removal of blocks from ten blocks that retain free radicals and prevents free flow of vatha or air.

ii. Physiological Information

Our conscious thinking and our somatic nerves are obviously controlled by our free will but our autonomic nerve system answers predominantly to a timing mechanism in our brain called the circadian clock. This function is primarily a response controlled by a section of our main gland of emotion (hypothalamus) called the supra-chiasmatic nucleus, which operates our sleep wake cycle.

The synchronization of this gland with nature is triggered by the Pineal gland, which interacts with several other glands for our large variety of timing functions. For example the Pons and Medulla interact with the Hypothalamus and the Pineal gland to regulate our breathing rhythm. The Substantia nigra of the midbrain which releases the hormone of love (Dopamine) interacts with the Hypothalamus and the Cortex to assist our 'stop watch timing capabilities'. This function enables us to mentally assess such things as when a cake is ready to come out of the oven or the arrival of a delivery that is due in one hour. Science has noted, removal of the pineal gland stops hormonal response to opposites of nature (day-night and summer-winter) which would bring about a need for constant hormone therapy.

(Neurosurgery (Vol 15 No6 1984.) P815 QUOTE.

"Removal of the pineal gland in a human resulted in a loss of the usual 24-hour rhythm of secretion."Science has also noted damage to the supra-chiasmatic nucleus also results in a loss of the 24-hour daily cycle and people who are blinded, loose their natural 24 hour biological rhythm as their eyes don't register the difference between day and night, to the Pineal gland. (The experience of twilight each evens and sunlight shining on our eyelids when the sun rises, triggers a release of the hormones that keep our body functions organized and in line with nature.)

The sleep improvement ideals are based around these biological clock functions, and tensions that interfere with hypothalamic efficiency, which in turn influences every mind body interaction.

"Focusing on goals is a necessity of our design" NoelBatten One of the needs of our system is for us to have goals to focus on to stimulate our forward movement in life. This is to ensure that Intelligence and wisdom is continually challenged and sharpened. If we were to tread water all the way through life we would learn nothing and would miss out on the excitements of achievement.

A. Malnutrition

Parkinson's disease can be termed as conditions created by malnutrition. Malnutrition also creates conditions like obesity and loss of weight. Dravidian cure corrects the injury to vital points in the body. Malnutrition induced by lactose intolerance can be corrected by appropriate change in the food habit or alternative medicine. It is doubtful whether modern medicine has any real medicine for lactose intolerance, gluten related injury. Most of the medicines given are for immune recovery suppression. Practitioners do not often disclose to the patients their limitations. Acute hunger pain is felt by even well fed patients. Symptoms of cure can be noticed by changes in the texture of skin. Perceptible improvement of the skin and health and formation of clear skin are indications of effectiveness of treatment in the early stages.

#

Parkinson's Disease Dravidian Cure Chintarmony System

Chapter IV

LOGO THERAPY

"The patient, brain washed into believing that there is no hope for him starts looking at life with hope."

Meaninglessness is a condition which the mind finds it hard to tolerate cannot tolerate It leads to boredom says Victor Frankl the psycho therapist. He who has a why to live can bear with almost any how, is the fundamental psyche. A person suffering from depressive psyche may behave like a Parkinson's patient. Such person also may be victim of dopamine intake. Such people need auto suggestion to keep mind under control.

Collective neurosis is a condition which is created by well orchestrated propaganda. Nazism, communism, many ideologies, god men and god women and idols were created to lure people and divert them from their normal life. When such movements fail, followers suffer psychological disorder. Romantic infatuations and failure also introduce depressions. Single term for such conditions is perhaps cognitive dissonance. Such depressed persons also suffer malnutrition. They also may become victim of dopamine administration. They suffer from acute anxiety and fear. One who looses strength of mind can die over night despite his absolutely healthy body. Parkinson's disease can also be caused by such anxiety, fear or shock. Fear is the mother of event. People pursuing faulty ideology suffer acute anxiety or depression. They loose confidence in life. Acute depression, fear anxiety causes uncontrollable stretching of involuntary muscles distorting vertebral column injuring the nerves. Chintarmony physician identifies nerve injury and heals the injured nerve.

1. Rational Emotive Behavioral Therapy

The REBT framework assumes that humans have both innate rational(meaning self- and social-helping and constructive) and irrational (meaning self- and social-defeating and un-helpful) tendencies and leanings. REBT claims that people to a large degree consciously and unconsciously construct emotional difficulties such as self-blame, self-pity, clinical anger, hurt, guilt, shame, depression and anxiety, and behaviors and behavior tendencies like procrastination, over-compulsiveness, avoidance, addiction and withdrawal by the means of their irrational and self-defeating thinking, emoting and behaving.[15] REBT is then applied as an educational process in which the therapist often actively teaches the client how to identify irrational and self-defeating beliefs and philosophies which in nature are rigid, extreme, unrealistic, illogical and absolutist, and then to forcefully and actively question and dispute them and replace them with more rational and self-helping ones. By using different cognitive, emotive and behavioral methods and activities, the client, together with help from the therapist and in homework exercises, can gain a more rational, self-helping and constructive rational way of thinking, emoting and behaving. One of the main objectives in REBT is to show the client that whenever unpleasant and unfortunate activating events occur in people's lives, they have a choice of making themselves feel healthily and self-helpingly sorry, disappointed, frustrated, and annoyed, or making themselves feel unhealthy and self-defeating, horrified, terrified, panicked, depressed, self-hating, and self-pitying.[16] By attaining and ingraining a more rational and self-constructive philosophy of themselves, others and the world, people often are more likely to behave and emote in more life-serving and adaptive ways.

Rational emotive behavior therapy (REBT), previously called rational therapy and rational emotive therapy, is a comprehensive, active-directive, philosophically and empirically based psychotherapy which focuses on resolving emotional and behavioral problems and disturbances and enabling people to lead happier and more fulfilling lives. REBT was created and developed by the American psychotherapist and psychologist Albert Ellis who was inspired by many of the teachings of Asian, Greek, Roman and modern philosophers. REBT is one form of cognitive behavior therapy (CBT) and was first expounded by Ellis in the mid-1950s; development continued until his death.

Rational Emotive Behavior Therapy (REBT) is both a psychotherapeutic system of theory and practices and a school of thought established by Albert Ellis. Originally called rational therapy, its appellation was revised to rational emotive therapy in 1959, then to its current appellation in 1992. REBT was one of the first of the cognitive behavior therapies, as it was predicated in articles Ellis first published in 1956,[7] nearly a decade before Aaron Beck first set forth his cognitive therapy. Precursors of certain fundamental aspects of REBT have been identified in various ancient philosophical traditions, particularly Stoicism. For example, Ellis' first major publication on rational therapy describes the philosophical basis of REBT as the principle that a person is rarely affected emotionally by outside things but rather by 'his perceptions, attitudes, or internalized sentences about outside things and events.' He adds,

This principle, which I have inducted from many psychotherapeutic sessions with scores of patients during the last several years, was originally discovered and stated by the ancient Stoic philosophers, especially Zeno of Citium (the founder of the school), Chrysippus most influential disciple], Panaetius of Rhodes (who introduced Stoicism into Rome), Cicero, Seneca, Epictetus, and Marcus Aurelius. The truths of Stoicism were perhaps best set forth by Epictetus, who in the first century A.D. wrote in the Enchiridion: "Men are disturbed not by things, but by the views which they take of them." Shakespeare, many centuries later, rephrased this thought in Hamlet: "There's nothing good or bad but thinking makes it so."

One of the fundamental premises of REBT is that humans, in most cases, do not merely get upset by unfortunate adversities, but also by how they construct their views of reality through their language, evaluative beliefs, meanings and philosophies about the world, themselves and others. This concept has been attributed as far back as the Greek Philosopher Epictetus, who is often cited as utilizing similar ideas in antiquity. In REBT, clients usually learn and begin to apply this premise by learning the A-B-C-model of psychological disturbance and change. The A-B-C model states that it normally is not merely an A, adversity (or activating event) that contributes to disturbed and dysfunctional emotional and behavioral Cs, consequences, but also what people B, believe about the A, adversity. A, adversity can be either an external situation or a thought or other kind of internal event, and it can refer to an event in the past, present, or future.

The Bs, beliefs that are most important in the A-B-C model are explicit and implicit philosophical meanings and assumptions about events, personal desires, and preferences. The Bs, beliefs that are most significant are highly evaluative and consist of interrelated and integrated cognitive, emotional and behavioral aspects and dimensions.

According to REBT, if a person's evaluative B, belief about the A, activating event is rigid, absolutistic and dysfunctional, the C, the emotional and behavioral consequence, is likely to be self-defeating and destructive. Alternatively, if a person's evaluative B, belief is preferential, flexible and constructive, the C, the emotional and behavioral consequence is likely to be self-helping and constructive.

Through REBT, by understanding the role of their mediating, evaluative and philosophically based illogical, unrealistic and self-defeating meanings, interpretations and assumptions in upset, people often can learn to identify them, begin to D, dispute, refute, challenge and question them, distinguish them from healthy constructs, and subscribe to more constructive and self-helping constructs. The REBT framework assumes that humans have both innate rational (meaning self- and social-helping and constructive) and irrational (meaning self- and social-defeating and un-helpful) tendencies and leanings. REBT claims that people to a large degree consciously and unconsciously construct emotional difficulties such as self-blame, self-pity, clinical anger, hurt, guilt, shame, depression and anxiety, and behaviors and behavior tendencies like procrastination, over-compulsiveness, avoidance, addiction and withdrawal by the means of their irrational and self-defeating thinking, emoting and behaving.[15] REBT is then applied as an educational process in which the therapist often actively teaches the client how to identify irrational and self-defeating beliefs and philosophies which in nature are rigid, extreme, unrealistic, illogical and absolutist, and then to forcefully and actively question and dispute them and replace them with more rational and self-helping ones. By using different cognitive, emotive and behavioral methods and activities, the client, together with help from the therapist and in homework exercises, can gain a more rational, self-helping and constructive rational way of thinking, emoting and behaving. One of the main objectives in REBT is to show the client that whenever unpleasant and unfortunate activating events occur in people's lives, they have a choice of making themselves feel health and self-helping sorry, disappointed, frustrated, and annoyed, or making themselves feel unhealthy and self-defeating, horrified, terrified, panicked, depressed, self-hating, and self-pitying.[16] By attaining and ingraining a more rational and self-constructive philosophy of themselves, others and the world, people often are more likely to behave and emote in more life-serving and adaptive ways.

Albert Ellis posits three major insights of REBT:

Parkinson's Disease Dravidian Cure Chintarmony System

Insight 1 - People seeing and accepting the reality that their emotional disturbances at point C only partially stem from the activating events or adversities at point A that precede C. Although A contributes to C, and although disturbed Cs (such as feelings of panic and depression) are much more likely to follow strong negative As(such as being assaulted or raped), than they are to follow weak As (such as being disliked by a stranger), the main or more direct cores of extreme and dysfunctional emotional disturbances (Cs) are people's irrational beliefs — the absolutistic musts and their accompanying inferences and attributions that people strongly believe about their undesirable activating events.

Insight 2 - No matter how, when, and why people acquire self-defeating or irrational beliefs (i.e. beliefs which are the main cause of their dysfunctional emotional-behavioral consequences), if they are disturbed in the present, they tend to keep holding these irrational beliefs and continue upsetting themselves with these thoughts. They do so not because they held them in the past, but because they still actively hold them in the present, though often unconsciously, while continuing to reaffirm their beliefs and act as if they are still valid.

Insight 3 - No matter how well they have achieved insight 1 and insight 2, insight alone will rarely enable people to undo their emotional disturbances. They may feel better when they know, or think they know, how they became disturbed - since insights can give the impression of being useful and curative. But, it is unlikely that they will actually get better and stay better unless they accept insights 1and 2, and then also go on to strongly apply insight 3: There is usually no way to get better and stay better but by: continual work and practice in looking for, and finding, one's core irrational beliefs; actively, energetically, and scientifically disputing them; replacing one's absolutist musts with flexible preferences; changing one's unhealthy feelings to healthy, self-helping emotions; and firmly

acting against one's dysfunctional fears and compulsions. Only by a combined cognitive, emotive, and behavioral, as well as a quite persistent and forceful attack on one's serious emotional problems, is one likely to significantly ameliorate or remove them — and keep them removed.

Parkinson's Disease Dravidian Cure Chintarmony System

Regarding cognitive-affective-behavioral processes in mental functioning and dysfunctioning, originator Albert Ellis explains:[16]

"REBT assumes that human thinking, emotion, and action are not really separate or disparate processes, but that they all significantly overlap and are rarely experienced in a pure state. Much of what we call emotion is neither more nor less than a certain kind — a biased, prejudiced, or strongly evaluative kind — of thought. But emotions and behaviors significantly influence and affect thinking, just as thinking influences emotions and behaviors. Evaluating is a fundamental characteristic of human organisms and seems to work in a kind of closed circuit with a feedback mechanism: First, perception biases response, and then response tends to bias subsequent perception. Also, prior perceptions appear to bias subsequent perceptions, and prior responses appear to bias subsequent responses. What we call feelings almost always have a pronounced evaluating or appraisal element."

Albert Ellis has suggested three core beliefs or philosophies that humans tend to disturb themselves through:

"I absolutely MUST, under practically all conditions and at all times, perform well (or outstandingly well) and win the approval (or complete love) of significant others. If I fail in these important—and sacred—respects, that is awful and I am a bad, incompetent, unworthy person, who will probably always fail and deserves to suffer." "Other people with whom I relate or associate, absolutely MUST, under practically all conditions and at all times, treat me nicely, considerately and fairly. Otherwise, it is terrible and they are rotten, bad, unworthy people who will always treat me badly and do not deserve a good life and should be severely punished for acting so abominably to me." "The conditions under which I live absolutely MUST, at practically all times, be favorable, safe, hassle-free, and quickly and easily enjoyable, and if they are not that way it's awful and horrible and I can't bear it. I can't ever enjoy myself at all. My life is impossible and hardly worth living." Holding this belief when faced with adversity tends to contribute to feelings of anxiety, panic, depression, despair, and worthlessness. Holding this belief when faced with adversity tends to contribute to feelings of anger, rage, fury, and vindictiveness. Holding this belief when faced with adversity tends to contribute to frustration and discomfort, intolerance, self-pity, anger, depression, and to behaviors such as procrastination, avoidance, and inaction.

REBT commonly posits that at the core of irrational beliefs there often are explicit or implicit rigid demands and commands, and that extreme derivatives like awful, frustration, intolerance, people deprecation and over-generalizations are accompanied by these.

According to REBT the core dysfunctional philosophies in a person's evaluative emotional and behavioral belief system, are also very likely to contribute to unrealistic, arbitrary and crooked inferences and distortions in thinking. REBT therefore first teaches that when people in an insensible and devout way overuse absolutistic, dogmatic and rigid "should(s)", "musts", and "ought(s)", they tend to disturb and upset themselves.

Further REBT generally posits that disturbed evaluations to a large degree occur through over-generalization, wherein people exaggerate and globalize events or traits, usually unwanted events or traits or behavior, out of context, while almost always ignoring the positive events or traits or behaviors. For example, awful is partly mental magnification of the importance of an unwanted situation to a catastrophe or horror, elevating the rating of something from bad to worse than it should be, to beyond totally bad, worse than bad to the intolerable and to a "holocaust". The same exaggeration and over generalizing occurs with human rating, wherein humans come to be arbitrarily and axiomatically defined by their perceived flaws or misdeeds. Frustration intolerance then occurs when a person perceives something to be too difficult, painful or tedious, and by doing so exaggerates these qualities beyond one's ability to cope with them.

Essential to REBT theory is also the concept of secondary disturbances which people sometimes construct on top of their primary disturbance.

"Because of their self-consciousness and their ability to think about their thinking, they can very easily disturb themselves about their disturbances and can also disturb themselves about their ineffective attempts to overcome their emotional disturbances."

As would be expected, REBT argues that mental wellness and mental health to a large degree results from an adequate amount of self-helping, flexible, logico-empirical ways of thinking, emoting and behaving.[15] When a perceived undesired and stressful activating event occurs, and the individual is interpreting, evaluating and reacting to the situation rationally and self-helpingly, then the resulting consequence is, according to REBT, likely to be more healthy, constructive and functional. This does not by any means mean that a relatively un-disturbed person never experiences negative feelings, but REBT does hope to keep debilitating and unhealthy emotions and subsequent self-defeating behavior to a minimum. To do this REBT generally promotes a flexible, un-dogmatic, self-helping and efficient belief system and constructive life philosophy about adversities and human desires and preferences.

That people had better accept life with its hassles and difficulties not always in accordance with their wants, while trying to change what they can change and live as elegantly as possible with what they cannot change without submitting to collective neurosis.

As explained, REBT is a therapeutic system of both theory and practices; generally one of the goals of REBT is to help clients see the ways in which they have learned how they often needlessly upset themselves, teach them how to un-upset themselves and then how to empower themselves to lead happier and more fulfilling lives.

2. Cognitive Dissonance

Cognitive dissonance is a discomfort caused by holding conflicting cognitions (e.g., ideas, beliefs, values, emotional reactions) simultaneously. In a state of dissonance, people may feel surprise, dread, guilt, anger, or embarrassment. The theory of cognitive dissonance in social psychology proposes that people have a motivational drive to reduce dissonance by altering existing cognitions, adding new ones to create a consistent belief system, or alternatively by reducing the importance of any one of the dissonant elements. An example of this would be the conflict between wanting to smoke and knowing that smoking is unhealthy; a person may try to change their feelings about the odds that they will actually suffer the consequences, or they might add the consonant element that the smoking is worth short term benefits. A general view of cognitive dissonance is when one is biased towards a certain decision even though other factors favor an alternative.

Disturbed sleep damages the vertebral column in the neck due to muscular stress/injury to the neck muscles. This was a common experience in West Bengal in India during Communist Rule. Funds allotted to roads repairs were fully diverted. Author was injured in a journey right from

Parkinson's Disease Dravidian Cure Chintarmony System

Calcutta to Darjeeling. Fortunately due to hectic activity and access to alternate remedies injury to vertebral column of the neck were ignored. Lack of the trust in the practice of modern medicine also helped author in avoiding allopathic doctors and suffer dopamine prescription. Author suffered steady loss of weight and faulty gait. On diagnosis by a Chintarmony traditional physician injuries to seven vertebral columns in the neck were diagnosed.

Inability of intestine and gall bladder to absorb calcium impedes vertebral column reconstruction due to calcium deficiency. The practice of surgeons to remove gall bladder at the slightest provocation has to be deprecated. If repair of roads are neglected, the traveling public are exposed to the risk of injured vertebral column of the neck and consequent Parkinson's disease. Mere jumping from a height can damage vertebral column joints damaging cartilage and nerves. Continuous auto immune disorders will lead to water retention in the body which can induce a range of disease from Parkinson disease to cancer.

#

Parkinson's Disease Dravidian Cure Chintarmony System

Chapter V

CAUSES OF PARKINSON'S DISEASE

A. Multiple Sclerosis

Modern medicine considered injury to central nervous system consisting of brain and spinal chord cannot be repaired once damaged. The condition can only be managed. Research is progressing and now considers that damage to central nervous system could be cured. Chintarmony system offers cure to the damaged central nervous system.

Multiple sclerosis (MS), also known as "disseminated sclerosis" or "encephalomyelitis disseminata", is an inflammatory condition in which the fatty myelin sheaths around the axons of the brain and spinal cord are damaged, leading to demylination and scarring as well as a broad spectrum of signs and symptoms. Disease onset usually occurs in young adults, and it is more common in women. It has a prevalence that ranges between 2 and 150 per 100,000.

MS affects the ability of nerve cells in the brain and spinal cord to communicate with each other effectively. Nerve cells communicate by sending electrical signals called action potentials down long fibers called axon, which are contained within an insulating substance called myelin. In MS, the bodies own immune response attacks and damage the myelin. When myelin is lost, the axons can no longer effectively conduct signals. The name multiple sclerosis refers to scars (scleroses—better known as plaques or lesions) particularly in the white matter of the brain and spinal cord, which is mainly composed of myelin. Although much is known about the mechanisms involved in the disease process, the cause remains unknown. Theories include genetic infections or. Different environmental risk factors have also been found.

Almost any neurological symptom can appear with the disease, and often progresses to physical and cognitive disability MS takes several forms, with new symptoms occurring either in discrete attacks (relapsing forms) or slowly accumulating over time (progressive forms). Between attacks, symptoms may go away completely, but permanent neurological problems often occur, especially as the disease advances.

Parkinson's Disease Dravidian Cure Chintarmony System

There is no known cure for multiple sclerosis in modern medicine though alternate medicines have convincing cure. Treatments attempt to return function after an attack, prevent new attacks, and prevent disability. Multiple sclerosis medications can have adverse effects or be poorly tolerated, and many patients pursue alternative treatments, despite the lack of supporting scientific study as understood in modern scientific jargon. The prognosis is difficult to predict; it depends on the subtype of the disease, the individual patient's disease characteristics, the initial symptoms and the degree of disability the person experiences as time advances. Life expectancy of people with multiple sclerosis is 5 to 10 years lower than that of the unaffected population.

1. Blood-Brain Barrier Breakdown

According to Chintharmony physician, it is not the ability of substrata nigra to produce dopamine neurons that causes the symptoms of Parkinson's disease. It is the inability of the neurons to break through the brain blood barrier due to accumulation of free radicals that leads to the condition. Chintarmony system induces electrons to remove free radicals through oil bath and massage.

Demyelination in MS. On Klüver-Barrera myelin staining, decoloration in the area of the lesion can be appreciated (Original scale 1:100).

The blood-brain barrier is a capillary system that should prevent entrance of T cells into the nervous system. The blood–brain barrier is normally not permeable to these types of cells, unless triggered by infection or a virus, which decreases the integrity of the tight junctions forming the barrier. When the blood–brain barrier regains its integrity, usually after infection or virus has cleared, the T cells are trapped inside the brain.

2. Auto Immunology

MS is currently believed to be an immune-mediated disorder mediated by a complex interaction of the individual's genetics and as yet unidentified environmental insults. Damage is believed to be caused by the patient's own immune system. The immune system attacks the nervous system, possibly as a result of exposure to a molecule with a similar structure to one of its own. Mw151 and Mw189 address accumulation of water in the brain. Brain cells are injured by the water pressure. But why there is accumulation of water is not probed.

3. Lesions

The name multiple sclerosis refers to the scars (scleroses – better known as plaques or lesions) that form in the nervous system. MS lesions most commonly involve white matter areas close to the ventricles of the cerebellum, brain stem, basal ganglia, and spinal cord; and the optic nerve. The function of white matter cells is to carry signals between grey matter areas, where the processing is done, and the rest of the body. The peripheral nervous system is rarely involved.

More specifically, MS destroys oligohydrocytes, the cells responsible for creating and maintaining a fatty layer—known as the myelin sheath—which helps the neurons carry electrical signals (action potentials). MS results in a thinning or complete loss of myelin and, as the disease advances, the cutting (transection) of the neuron's extensions or axons. When the myelin is lost, a neuron can no longer effectively conduct electrical signals. A repair process, called remyelination, takes place in early phases of the disease, but the oligodendrocytes cannot completely rebuild the cell's myelin sheath. Repeated attacks lead to successively fewer effective remyelinations, until a scar-like plaque is built up around the damaged axons. Different lesion patterns have been described.

4. Inflammation

Apart from demyelination, the other pathologic hallmark of the disease is inflammation. According to a strictly immunological explanation of MS, the inflammatory process is caused by T cells, a kind of lymphocyte. Lymphocytes are cells that play an important role in the body's defenses. In MS, T cells gain entry into the brain via the previously described blood-brain barrier. Evidence from animal models also point to a role of B cells in addition to T cells in development of the disease.

The T cells recognize myelin as foreign and attack it as if it were an invading virus. This triggers inflammatory processes, stimulating other immune cells and soluble factors like cytokines and antibodies. Leaks form in the blood–brain barrier, which in turn cause a number of other damaging effects such as swelling, activation of macrophages, and more activation of cytokines and other destructive proteins.

Cure of Pd by chintarmony physician is by correcting the brain blood barrier and repairs demyelinised nerve sheath and with mudra kriya dopamine neurons are helped to flow in to nerve centers in the body. This could be offered as a scientific basis of the cure. Chintarmony physician follow a structured path with out offering scientific jargon. The cat always caught the rat in the system. Physicians are prohibited from experimenting with the patients. They strictly adhere to the tradition.

B. Salt Deficiency

Sodium and potassium deficiency can induce Parkinson's disease like symptom leading to dopamine trap.

Refined Salt: White Poison

The problem with salt is not the salt itself but the condition of the salt we eat! Our regular table salt no longer has anything in common with the original crystal salt. Salt now a day is mainly sodium chloride and not salt. With the advent of industrial development, our natural salt was "chemically cleaned" and reduced only to sodium and chloride. Major producing companies dry their salt in huge kilns with temperatures reaching 1200 degrees F, changing the salt's chemical structure, which in turn adversely affects the human body. The common table salt we use for cooking has only 2 or 3 chemical elements. The seawater has 84 chemical elements. For our body to be healthy we need all those elements. When we use the common salt, we are in deficit of 81 elements which means we are somehow contributing to becoming weaker, imbalanced and more susceptible to diseases. Use the seawater salt.

C. Food Culture

Fermented rice soup is a traditional food used in Kerala, India. The fermented food enhances liver function by decreasing digestive load. This food can be applied for a variety of conditions like, multiple sclerosis, bacterial imbalance, liver malfunction gluten related injury etc.

Parkinson's Disease Dravidian Cure Chintarmony System

Gluten a sticky protein is a major culprit in diseases ranging from Parkinson's disease to thyroid, blood pressure, cardiac conditions, spondilosis, cholesterol etc. Gluten is a composition of two proteins gluten in and gliding. Gluten sticks to the villi in the small intestine preventing alkaline production. Villi look like outer covering of jackfruit. Gluten deeply stick to the mucus membrane covering villi, immense system perceives the situation as an aggression by a foreign body. Immune symptom is not able differentiate between sticky gluten and mucus covering. Entire mucus line gets destroyed ulcerating the intestinal lining. Yeast and other foreign body penetrate through these wounds and colonize the whole body.

The body has to fight continuously and soon gets exhausted. The condition is described as immune recovery suppression. Intermittent fevers are only efforts in immune recovery for steady long war leading to water retention in the body destroying muscular balance. Muscles sag. Posture of the patient also changes. Soon patient suffers from cartilage damage causing damages to nerves. Finally body experiences a condition called immune recovery suppression which is now labeled as AIDS or auto immune disorder. The body is not able to recoup and fight foreign bodies and surrenders to even the weakest infections. The patient will be tested AIDS positive. Dravidian cures term cancer as a condition caused by water retention and after simple, remedies. Biopsy is considered by them as a fundamentally faulty approach driving often the patient to a point of no return. According to Dravidian cure, biopsy triggers spread of dormant carcinogenic cells. Piles itself is termed in the ancient system as a form of cancer. Nobel laurite Linus Pauling offers simple remedy for cancer and heart ailments by intravenous injection of vitamin C. Linus Pauling and Dr.Rath Mathias convincingly proved that simple inexpensive remedy is possible for cancer, heart disease and AIDS. Medical evidence is clear that statin tablets increases the chance of heart attack. Modern medicine trails much behind 5000 B.C. in many areas where Dravidian cure excels.

D.   Fear Factor

Fear, anxiety and tension induce digestive disorders. Many rustic healers in India practicing Dravidian Cures consider Parkinson's disease as fear related disease. Involuntary muscles are controlled by sub conscious mind. Fear act as a fierce force stretching involuntary muscles beyond the control of conscious mind. Acute damage to the vertebral column is caused by such intense fear.

Parkinson's Disease Dravidian Cure Chintarmony System

Parents and teachers managing children with fear than love are destroying the body and mind of children. Adults are also encounter fear. Counseling and confidence building exercise for psychological healing is also needed.

E.   Postures

Patient in progression of disease behaves like a Famine stricken community. All over the world famine stricken people and Parkinson's patient walks the same way. Stiff neck and vertebral column, unsteady walk and protruding eyes are the common traits of Parkinson's patient. Chintamoni system terms these behaviour as different facets of Dasa Vasthas ie, ten points where flow of energy gets blocked due to oxidation stress, ie vatha. The terminal stage is Vatha hitting brain and eyes after blocking the energy flow through limbs to Vertebral column. Noelbatten of Australia terms Parkinson's disease as inadequate oxygen content during R.E.M. sleep. Dravidian cure terms the injury as due to oxidative stress. Medicines in Dravidian cure are positive charged and have antioxidant properties only a person on vegetarian diet is allowed to collect plants to preserve positive charge. Metal knife cannot be used to cut plants to preserve charge of plants. They follow super science evolved around 5000 B.C. Cure is concluded by correction of vertebral column injury by application of oils and massage.

F. Lactose Intolerance

Yet another cause of Parkinson's disease is lactose intolerance, which induces malnutrition. Muscular arrangement is lost in such patients leading to vertebral column misalignment.

When lactose moves through the large intestine colon without being properly digested, it can cause uncomfortable symptoms such as gas, belly pain, and bloating. Some people with lactose intolerance cannot digest any milk products. Others can eat or drink small amounts of milk products or certain types of milk products without problems.

Lactose intolerance is common in adults. It occurs more often in Native Americans and people of Asian, African, and South American descent than among people of European descent.

Parkinson's Disease Dravidian Cure Chintarmony System

A big challenge for people who are lactose-intolerant is learning how to eat to avoid discomfort and to get enough calcium for healthy bones. Lactose intolerance occurs when the large intestine does not make enough of enzyme called lactase. Body needs lactase to break down, or digest, lactose.

Lactose intolerance most commonly runs in families, and symptoms usually develop during the teen or adult years. Most people with this type of lactose intolerance can eat some milk or dairy products without problems.

Sometimes the small intestine stops making lactase after a short-term illness such as the stomach flu or as part of a lifelong disease such as cystic fibrosis. Or the small intestine sometimes stops making lactase after surgery to remove a part of the small intestine. In these cases, the problem can be either permanent or temporary.

In rare cases, newborns are lactose-intolerant. A person born with lactose intolerance cannot eat or drink anything with lactose.

Some premature babies have temporary lactose intolerance because they are not yet able to make lactase. After a baby begins to make lactase, the condition typically goes away.

Symptoms of lactose intolerance can be mild to severe, depending on how much lactase your body makes. Symptoms usually begin 30 minutes to 2 hours after you eat or drink milk products. If you have lactose intolerance, your symptoms may include bloating, Pain or cramps, gurgling or rumbling sounds in your belly, gas, loose stools or diarrhea, throwing up.

Many people who have gas, belly pain, bloating, and diarrhea suspect they may be lactose-intolerant. The best way to check this is to avoid eating all milk and dairy products to see if your symptoms go away. If they do, then you can try adding small amounts of milk products to see if your symptoms come back.

If you feel sick after drinking a glass of milk one time, you probably do not have lactose intolerance. But if you feel sick every time you have milk, ice cream, or another dairy product, you may have lactose intolerance.

F. GLUTEN RELATED INJURY

GRI could be another cause of Parkinson's disease, induced by malnutrition due to acute villi damage.

Gluten is a protein found in the grains of several grass crops, including wheat, rye, barley and their relatives. It is widely used in the food industry as a thickener, stabilizer and filler. Gluten is only partially digested in your intestine, giving rise to protein fragments that can trigger a robust immune response in individuals whose genetic makeup makes them sensitive to gluten. Celiac disease, which is characterized by elevated antibodies to gluten, intestinal damage and abnormalities in various other organs, is a well-recognized gluten-sensitivity disorder. Your nervous system is one of the organ systems that can be adversely affected by gluten.

1. Wide Ranging Injury

Physicians have known for years that celiac disease can affect tissues outside your gastrointestinal tract. Gluten sensitivity has been linked to disorders of the skin, liver, skeleton, endocrine organs and reproductive system. Neurologic injury associated with celiac disease was first reported in 1966. It has generally been assumed that involvement of these organ systems occurred in conjunction with intestinal damage. However, as scientists learn more about gluten's effects on your body, it is becoming apparent that gluten sensitivity can disrupt the function of many tissues without necessarily causing intestinal symptoms.

2. Ataxia

In the February 2011 issue of "Diagnostic Pathology," scientists at Tokyo Medical University in Tokyo, Japan, reported the case of an 84-year-old woman who developed ataxia – loss of coordination and balance – due to gluten sensitivity. This patient's symptoms mirrored those of other patients with gluten-sensitive ataxia, several of whom died from non gluten-related causes shortly after their diagnoses were made. Post-mortem examination of these individuals' brains revealed inflammation and destruction of brain cells that help to control balance and movement. Often such patients are also given dopamine. New generation medicine MW151 and Mw189 addresses inflammation by dehydrating cells by inducing immuno recovery suppression.

Parkinson's Disease Dravidian Cure Chintarmony System

The medicine apparently is not able to cure the basic cause of water retention compelling continued life long dependence on the medicine. Water pressure killing the brain is the root cause of epilepsy, rabies, multiple sclerosis, Parkinson's disease, arthritis etc

3. Hearing Problems

Celiac disease has been linked to hearing abnormalities in pediatric patients, and it probably affects adults, as well. A 2011 study at Ankara, Turkey's MH Kecioren Training and Research Hospital demonstrated hearing loss and increased "background noise" in 41 children with celiac disease, and a 2007 trial published in the Polish journal "Otolaryngologia Polska" revealed that gluten-related damage to the inner ear persisted in children aged 6 to 18 years despite adherence to a gluten-free diet.

4. Cognitive Decline

A review of records from patients referred to the Mayo Clinic from 1970 to 2005 identified 13 individuals who developed progressive cognitive decline -- memory loss, difficulties with calculations, confusion and personality changes -- in association with celiac disease. In more than a third of these patients, their mental disturbances developed simultaneously with the onset of celiac disease. Several of these individuals exhibited significant improvement in their symptoms once they adopted a gluten-free diet.

5. Expanding Horizons

Gluten sensitivity encompasses a spectrum of disorders that range from mild gastrointestinal symptoms through isolated impairment of specific organ systems to a multisystem autoimmune disease. Nervous system involvement includes ataxia, hearing impairment, cognitive decline, seizures and peripheral neuropathies. Evidence suggests that some cases of attention-deficit disorder, learning disabilities, tic disorders, headaches and even schizophrenia could have their origins in gluten sensitivity. As scientists learn more about gluten's effects on your body, the widespread use of gluten in our food supply might fall under closer scrutiny.

The challenge that remains to be addressed is to discover how these neurons are destroyed to cause Parkinson's disease.

Parkinson's Disease Dravidian Cure Chintarmony System

6. Theories

Water retention in the body and muscular fatigue due to various reasons could be yet another cause. Many theories have been put forward, but most researchers believe that Parkinson's disease is not due to a single culprit but rather a combination of both genetic susceptibility and environmental stresses causing brain cell death.

Brain cell death is not an acceptable theory for chintarmony physicians. Brain cell death is a final phase of Parkinson's disease progressing through limbs, vertebral column and brain due to free radicals from oxidative stress.

Studies have found that living in a rural area, drinking well water, or being exposed to pesticides, herbicides, or wood pulp mills may increase your risk for developing Parkinson's disease due to sagging of muscle and vertebral column misalignment causing injury to nerves disrupting digestion or absorption or obstructing brain blood barrier.

It has been demonstrated that 5-10% of people with PD have a genetic tendency. A recent study identified a specific gene mutation in a group of people who were related. Although this gene mutation is not responsible for all causes of PD, this finding may give scientists the opportunity to develop an animal model to gain insight into PD.

Currently, one of the most promising theories is the oxidation hypothesis which is the fundamental on which chintarmony physician excels over all other system of cure.

It is thought that free radicals may play a role in the development of Parkinson's disease. Free radicals are chemical compounds with a positive charge that are created when dopamine is broken down by combining it with oxygen. This breakdown of dopamine by an enzyme called monoamine oxidase (MAO) leads to the formation of hydrogen peroxide peroxide.

A protein called glutathione normally breaks down the hydrogen peroxide quickly. If the hydrogen peroxide is not broken down correctly, it may lead to the formation of these free radicals that then can react with cell membranes to cause cell damage and something called lipid peroxidation (when the hydrogen peroxide interacts with lipids [fat soluble substances] in the cell membrane). In PD, glutathione is reduced, which may mean that you have a loss of protection against the formation of these free radicals.

Parkinson's Disease Dravidian Cure Chintarmony System

Also, iron is increased in the brain and may help form free radicals. In addition, lipid peroxidation is increased in Parkinson's disease. The association of Parkinson's disease with increased dopamine turnover, decreased mechanisms (glutathione) to protect against free radical formation, increased iron (which makes it easier to create free radicals), and increased lipid peroxidation helps support the oxidation hypothesis.

If this hypothesis turns out to be correct, it still does not explain why or how a loss of the protective mechanism occurs. An answer to this question may not be required. If the theory is correct, drugs may be developed to stop or delay these events. Although the cause of Parkinson's disease is not known, some people have symptoms of PD that may have an identifiable cause. In this case, the syndrome is known as parkinsonism or secondary PD.

It is thought that although primary Parkinson's, or Parkinson's disease, is the most common type seen by neurologists, Parkinson's that is caused by drugs is probably far more common than reported and accounts for about 4% of all cases of Parkinson's. A change in the level of dopamine, whether by brain cell loss or drug use, can create the symptoms of PD.

Interestingly, people who experience drug-induced Parkinson's may actually have a higher risk of developing PD later in life. A number of medications can cause Parkinson's by lowering dopamine levels. These are referred to as dopamine receptor antagonists or blockers.

Nearly all antipsychotic or neuroleptic medications such as chlorpropamazine (Thorazine), haloperidol (Haldol), and thioridazine (Mellaril) can induce the symptoms of Parkinson's. The medication valproic acid valporic acid (Depakote), a widely use antiseizure medication, may also cause a reversible Parkinson's.

Medications such as metoclopramide (Octamide, Maxolon, Reglan), which is used to treat certain stomach disorders such as peptic ulcers disease, are capable of causing Parkinson's or making it worse. Antidepressants known as selective serotonin-reuptake inhibitors may cause symptoms similar to Parkinson's. Central to all these medications is their ability to alter the concentration of dopamine in the central nervous system. Therefore a careful review of a medication list and ruling out of other causes such as brain tumors, stroke, infections, toxins, AIDS, and hydrocephalus must occur before the absolute diagnosis of Parkinson's disease is made.

7. Bacterial Balance

Bacterial balance is yet another circumstance which can induce digestive disorder. Around five pounds of beneficial bacteria is needed to maintain a symbiotic bacterial balance to assist healthy absorption and digestion. Environmental pollution, abuse of antibiotic, chlorinated water etc destroys bacterial yeast balance. Such imbalance causes malnutrition and water retention. According to ancient Indian pharmacopeia water retention is the cause of cancer.

#

Parkinson's Disease Dravidian Cure Chintarmony System

Chapter -VI

SIDHA CULT

Siddha is also a traditional medical system of India. It is of Dravidian origin and has its entire literature in Tamil language. The basic concepts of the Siddha medicine are the same as those of Ayurveda. The difference is mostly in detail, Siddha being influenced by the local tradition with roots in the ancient Dravidian culture.

Its origin is also traced to mythological sources belonging to the Shaiva tradition. According to the tradition, Lord Shiva conveyed the knowledge of medicine to his wife Parvati. The knowledge was passed from her to Nandi and finally it was given to the Siddhas. The word Siddha denotes one who has achieved some extraordinary powers (siddhi). This achievement was related to the discipline of mind and its superiority over body, and was accomplished through both yoga and medicine. Thus siddhars (practitioners of Siddha) became the symbols of psychosomatic perfection and so the Siddha medicine became a combination of medicine and yoga.

The tantrik siddhi was thought of in different forms such as janmaja (due to birth), osadhija (due to some medical elixirs), mantraja (due to magical incantations), tapoja (due to penance) and samadhija (due to meditation). The tantriks endeavoured to attain the siddhis by several means, one of them was through the use of certain compositions of compounds of mercury, sulphur, mica and several other metallic substances.

According to tradition, there were 18 Siddhars (the person who has achieved some extra-ordinary powers): Nandi, Agasthiyar, Thirumular, Punnakkeesar, Pulasthiyar, Poonaikannar, Idaikkadar, Bogar, Pulikai isar, Karuvurar, Konkanavar, Kalangi, Sattainathar, Azhuganni, Agappai, Pumbatti, Theraiyar and Kudhambai, but the Agasthiyar (Agastya) was the topmost. He is regarded as the originator of the Siddha medicine and also of the Tamil language. He occupies the same position as Hippocrates in modern western medicine. In the period of Ramayana he seems to have settled in the South. Thus origin of every tradition in the South, including language and culture, is traced back to Agastya. In the Siddha medicine system use of metals, minerals and chemical products is predominant. The use of metals started from the period of Vagbhata (6th Centaury AD). Alchemy actually has its origin in the Siddha system which was connected with the Tantrik culture, aimed at perfection of man not only at the spiritual level but also at the physical level. The use of human urine in medicine also started with the Tantrik culture and became popular in the medieval period.

The dates of most of alchemy texts are generally uncertain, but they belong possibly to a period between the 9th and the 18th Centuries AD, the period between the 10th and the 14th Centuries being perhaps the most flourishing one. Generally these texts come under the category of the rasasastra, signifying systematic treatments of the new knowledge and practices relating to the use of mercurial compounds and a host of other substance as medicine. The following are among the important rasasastra texts in Sanskrit: Rasahrdaya by Govinda Bhagavat, Rasaratnakara by Siddha Nagarjuna, Rasarnava (author unknown), Rasaratnasamuccya by Vagbhata, Rasaratnakara by Nityanatha Siddha, etc.

There are also some tantrik texts, which deal with alchemical ideas as part of their psycho-experimental-symbolic treatment for the tantrik goals and related practice. These texts are not only in Sanskrit language but also in other languages like Tamil, Telugu, Kannada, etc. About two hundred works in Tamil on the Siddha medicine having alchemical ideas. Of special importance are Amudakalaijnanam, Muppu, Muppuvaippu, Muppucunnam, Carakku, Guruseynir, Paccaivettusutram and Pannir-kandam by Agastya; Kadaikandam, Valalai-sutram and Nadukandam by Konganavar; Karagappa, Purva, Muppu-sutram and Dravakam by Nandisvar; Karpam and Valai-sutram by Bogar etc. The name of Agastya and Bogar have been mentioned as the authors of alchemy works in Tamil language. The writings of Bogar contain a number of references to his contacts with China. Whether he was a Chinese who imparted alchemical knowledge to the Tamilians is a moot point.

The alchemical literature in Sanskrit is presented as a dialogue between Siva and Parvati in their different forms, of which perhaps the most significant are the forms of Bhairava and Bhairavi. Siva is also worshipped in the form of known as linga. In Tamil language lingam also means cinnabar (mercuric sulphide) also, and that cinnabar forms one of the constituents of a composition (astabandha) used during the installation of divine idols. Traditionally cinnabar is the source of divine energy and possesses the creative principles.

Parkinson's Disease Dravidian Cure Chintarmony System

One of the Siddhars of Tamilnadu, Ramadevar, says in his work on alchemy (Cunnakandam) that he went to Mecca, assumed the name of Yakub and taught the Arabs the alchemical arts. It is significant that some of the purification processes and substances of alchemical significance are common to both Islamic and Indian alchemy. This perhaps establishes the theory that the sidha tradition originated from tribes in Africa

#

Parkinson's Disease Dravidian Cure Chintarmony System

Chapter-VII

BASICS OF SIDDHA MEDICINE

Generally the basic concepts of the Siddha medicine are almost similar to Ayurveda. The only difference appears to be that the Siddha medicine recognizes predominance of vatham, pitham and kapam in childhood, adulthood and old age respectively, whereas in Ayurveda it is totally reversed: kapam is dominant in childhood, vatham in old age and pitham in adults.

According to the Siddha medicine various psychological and physiological functions of the body are attributed to the combination of seven elements: first is saram (plasma) responsible for growth, development and nourishment; second is cheneer (blood) responsible for nourishing muscles, imparting colour and improving intellect; the third is ooun (muscle) responsible for shape of the body; fourth is kollzuppu (fatty tissue) responsible for oil balance and lubricating joints; fifth is enbu (bone) responsible for body structure and posture and movement; sixth is moolai (nerve) responsible for strength; and the last is sukila (semen) responsible for reproduction. Like in Ayurveda, in Siddha medicine also the physiological components of the human beings are classified as Vatha (air), Pitha (fire) and Kapha (earth and water).

1. **Concept of Disease and Cause**

When the normal equilibrium of three humors (vatha, pitha and kapha) is disturbed, disease is caused. The factors, which affect this equilibrium are environment, climatic conditions, diet, physical activities, and stress. Under normal conditions, the ratio between these three humors (vatha, pitha and kapha) is 4:2:1 respectively.

According to the Siddha medicine system diet and life style play a major role not only in health but also in curing diseases. This concept of the Siddha medicine is termed as pathya and apathya, which is essentially a list of do's and dont's.

2. **Method of Diagnosis**

In diagnosis, examination of eight items is required which is commonly known as astasthana-pariksa.

These are:

1. na (tongue): black in vatha, yellow or red in pitha, white in kapha, ulcerated in anaemia.
2. varna (colour): dark in vatha, yellow or red in pitha, pale in kapha;
3. svara (voice): normal in vatha, high pitched in pitha, low pitched in kapha, slurred in alcoholism.
4. kan (eyes): muddy conjunctiva, yellowish or red in pitha, pale in kapha.
5. sparisam (touch): dry in vatha, warm in pitha, chill in kapha, sweating in different parts of the body.
6. mala (stool): black stools indicate vatha, yellow pitha, pale in kapha, dark red in ulcer and shining in terminal illness.
7. neer (urine): early morning urine is examined; straw colour indicates indigestion, reddish yellow excessive heat, rose in blood pressure, saffron colour in jaundice and looks like meat washed water in renal disease.
8. nadi (pulse): the confirmatory method recorded on the radial artery.

3. Concept of Medicine

In Siddha medicine the use of metals and minerals are more predominant in comparison to other Indian traditional medicine systems. In the usage of metals, minerals and other chemicals, this system was far more advanced than Ayurveda. Siddhar Nagarjuna introduced the use of mercury and its compounds to the Ayurvedic system in later periods. The use of more metals and chemicals was justified by the fact that to preserve the body from decomposing materials that do not decompose easily should be used. The other reason perhaps was that the south Indian rivers were not perennial and herbs were not available all through the year.

The drugs used by the Siddhars could be classified into three groups: thavaram (herbal product), thathu (inorganic substances) and jangamam (animal products). The thathu drugs are further classified as uppu (water soluble inorganic substances or drugs that give out vapour when put into fire), pashanam (drugs not dissolved in water but emit vapour when fired), uparasam (similar to pashanam but differ in action), loham (not dissolved in water but melt when fired), rasam (drugs which are soft) and ghandhagam (drugs which are insoluble in water, like sulphor).

Parkinson's Disease Dravidian Cure Chintarmony System

- In herbal drugs, the Siddhars not only used herbs, which grow in the surrounding areas, but also herbs that grow in high altitudes of Himalayas. It is noteworthy that Siddhar Korakkar was the first to introduce Cannabis as a medicine; he used it as a powerful painkiller. They also used animal products as medicine, for example in mental diseases, peranda bhasma is used which is made of human skull bones and the skulls of dogs.
- The drugs used in Siddha medicine were classified on the basis of five properties: suvai (taste), guna (character), veerya (potency), pirivu (class) and mahimai (action).

According to their mode of application the Siddha medicine could be categorized into two classes: (1) internal medicine and (2) external medicine.

- Internal medicine was used through the oral route and further classified in to 32 categories based on their form, methods of preparation, shelf life, etc.
- External medicine includes certain forms of drugs and also certain applications like nasal, eye and ear drops and also certain procedures like leech application.
- According to their pharmaceutical preparations, Siddha medicine could be categorized into:
- Kudineer churanam (decoction powder): It is a fine powder of drugs.
- Chendooram: It is a red colour powder generally made of metallic compounds.
- Chunnam: It is alkaline in nature.
- Kalangu: It is based on mercury.
- Karpams: It could be herbal or non-herbal in nature, made on a daily basis.
- Karruppu: Mercury and sulphur are essentially present and its colour is dark black.
- Legiyams and rasayanams: It contains ghee, honey and sugar, apart from herbal powder and juices.
- Mathirai and vadagam: It is pills prepared from fine powdered paste.
- Maappagu: It is flavoured medicinal syrup and contains generally aromatic herbs, honey and sugar.
- Mezhugu, kuzhambu, kalimbu and mai: All of these categories have a waxy feel.
- Ney: It is medicated ghee, which contains fat-soluble plant substances.
- Pakkuvam and theenooral: It is herbal medicine with honey.

Parkinson's Disease Dravidian Cure Chintarmony System

- Parpam: It is prepared by the process of calcination.
- Patangam: It contains mercury, camphor, etc.
- Thailam: It is medicated oil; usually sesame seed oil, coconut oil, castor oil, etc are used in its preparation.
- Theeneer: It is distilled essence, which contains volatile constituents of the drugs.
- Mercury is used in five forms such as rasam (mercury), lingam (red sulphide of mercury), veram (mercury perchloride), pooram (mercury subchloride) and rasa-chinduram (red oxide of mercury). They are known as panchasutha.

In addition to drugs, pranayama and other disciplines of yoga are necessary for good health and longevity.

4. Concept of Treatment

The treatment in Siddha medicine is aimed at keeping the three humors in equilibrium and maintenance of seven elements. Proper diet, medicine and a disciplined regimen of life are advised for a healthy living and to restore equilibrium of humors in diseased condition. Saint Thiruvalluvar explains four requisites of successful treatment. These are the patient, the attendant, physician and medicine. When the physician is well qualified and the other agents possess the necessary qualities, even severe diseases can be cured easily. The treatment should be commenced as early as possible after assessing the course and cause of the disease. Treatment is classified into three categories: devamaruthuvum (Divine method); manuda maruthuvum (rational method); and asura maruthuvum (surgical method). In Divine method medicines like parpam, chendooram, guru, kuligai made of mercury, sulphur and pashanams are used. In the rational method, medicines made of herbs like churanam, kudineer, vadagam are used. In surgical method, incision, excision, heat application, blood letting, leech application are used.

According to therapies the treatments of Siddha medicines could be further categorized into following categories such as Purgative therapy, Emetic therapy, Fasting therapy, Steam therapy, Oleation therapy, Physical therapy, Solar therapy and Blood letting therapy, Yoga therapy, etc. This timeless medical system has two branches—Siddha massage and Siddha medicine. But the treatment as a whole works on removing energy blocks from the ten vital points of prana or energy concentration. Treatment also includes 13 retrieval systems, which have cured problems

Parkinson's Disease Dravidian Cure Chintarmony System

like cervical spondylitis, skeleto-muscular pains, rheumatic lumbar, back pain, slip disc and even paralysis!

5. Concept of Physician

In Siddha system of medicine a physician should be spiritual and have an in-depth knowledge about normal/abnormal functioning of the three humors, capable of curing ailments, intelligent, truthful, confident, associated with the elite, capable of preparing high quality drugs with mastery over medical classes. According to Theraiyar (a siddha) in his Thylavarga churrukama, the physician should have pure thought and action, love for all human beings, a detailed knowledge about geographical seasonal variations, correct physical and mental state and dietary habits. Agasthiyar Sillaraikkovai further adds generosity, patience, untiring hard work, capability of overcoming greed, anger, knowledge about astrology and numerology as the qualities of a physician. He says that a physician should protect his patient like an eyelid, which protects the eyes and care as a mother who cares for her sick child.

A physician should not wear colourful dress, nor use silk, leather rope, cosmetics and should always move around in white dress, using only sandal paste as cosmetics. Theraiyar in his Thylavarka churukkam insists that a physician should clean his hands many times and have bath after examining a patient.

6. Approaches

Modern medical practitioners once proudly carried themselves in deliberately unclean hands and attire to make them look busy practitioner. Unclean hands killed many patients even after exposing the faulty practices,despite the scientific pretensions of the age. Greed in doctors was considered a sin in the sidha tradition. Sidha tradition gives a dress code. Simple dress pleasing to the patient is advised. Physician has to wear clean dress and clean and hygiene body when attending the patients in the ancient sidha tradition.

Mental attitude of sidha tradition is to give expeditious cure. Approach of the modern medicine to avoid simple expeditious cure to pursue expensive and degrading cure is not followed by sidha tradition

7. Varma Branch Of Siddha Medicine

This branch of Siddha medicine that is being practiced in pockets of Tamil Nadu and Kerala is called Varma and is also called as marmani. This branch of science deals more with traumatology and accidental injuries than the internal injuries where no immediate symptoms are visibly seen. There are about a hundred vital points, which are either junctions of bones, tendons or ligaments or blood vessels, and are called varma points.

The concepts of Siddha medicine system are similar to Ayurveda, but in the Siddha medicine the use of metal and minerals is predominant. Pulse reading and urine testing are important features of the Siddha medicine. Pulse reading was considerably developed by the Siddhas and was used in diagnosis and prognosis of diseases. Putting oil drops on the surface of urine and observing their movement was used to conduct urine examination. Besides, smell, colour, deposits, etc are also observed. Thus the Siddha system is not basically a regional variant of Ayurveda conditioned by the local Tamil culture and tradition. Certain others trace the roots of sidha medicine to Dravidian migrants from Africa.

#

Parkinson's Disease Dravidian Cure Chintarmony System

Chapter- VIII

ACTION STEP

Get yourself diagnosed by a Chintarmony physician in India. Take holiday to State of Kerala an exotic location in South India. Spend two weeks in absolute calm and comfort. The state is known for exceptional streak climate ranging from 16* to 33* Celsius .In around 150 Kms state has sea coast, middle ranges and high ranges touching 5000ft. May be minor ailments like ARI , acute sinusitis also can be corrected. Stroke and cardiovascular disease are often corrected in the Chintarmony system in advance of the impending disaster.

CONCLUSION

The above discussion shows a simpler and logical approach is made by Chintarmony system of cure. Looking for complicated expensive cure is now the medical practice. More and more physicians should be trained in the system.

Tendency to preserve the system in the traditional family has caused drain of many valuable remedies.

One need only take 6 following steps towards cure,

1. Parkinson's disease is curable. Mind is a powerful mechanism. If mind fails nothing can save the patient. Demoralising the patient is only to submit the patients to pharmaceutical exploitation.
2. Reject the doctor who says Parkinson's disease is incurable and prescribe dopamine tablets. As unscrupulous doctors is on the increase double check your physician.
3. After consulting a homeopath take 3 drops of Ignatia 30 to restore R.E.M sleep. Ignatia puts patient to a happy mood.
4. The cure may not be satisfactory if the patient is too aged, abused by long dopamine medication, nerves are snapped beyond repair. There are too many variables in the body which may interrupt with absolute cure.
5. Try gooseberry juice mixed with aloe vera juice improving dose from 5ml to bowel tolerance level or intravenous administration of Vitamin C.

6. Meet a chiropractor and a counselor to fix body and mind before meeting seeking Chintarmony treatment

   Contact author V.Jayaprasad , vasanth beat no.3, manikiri road, kochi, state of Kerala India to route the patients to experts in South India.

E-mail Address: vasu.jayaprasad at gmail.com   91, 9846594136

DISCLAIMER

The book is intended to bring to the notice of Parkinson's patient various views to give hope and restoring confidence and a healthy mental aptitude to the patient. Only after examination of the pulse by chintarmony physician a cure can be guaranteed. Cure often depends on the extent of damage to energy flow. Author is a practicing lawyer and has no claim of superior medical knowledge. Attempt is to piece together scattered information to enable modern and alternate medicines to coexist and share information. Comparative study of medical practice should be part of syllabus in medical colleges.

#V#